Mind-blowing Ways How To Make Money Online In 2020

"Bootstrap Your Way To 20K Monthly"

By

B J Carter

All content ©Copyright material and published by Uply Media, Inc 2020. All rights reserved. This material can not be reproduced without permission.

Contents

Introduction

Session 1: How-to Create Information Products That Sell

Session 2: Dot Crypto Blockchain Domain Investing and Flipping

Session 3: Master Affiliate Marketing For Six-Figure Income

Session 4: Secret Sauce To Selling Massive Kindle Downloads

Bonus Secrets!

Introduction

"So you finally decided it's time to start an online business??"

About time, *"what took you so long?!"*

Congratulations, you made a good choice by selecting the online business revenue model.

Chances are you have already researched a plethora of online business opportunities, now you ready to make a move. However, you need some guidance on which online business will make you the most money the quickest and how to get started on a bootstrap budget.

"The good news is that you can make 20k or more monthly with an online business!"

The bad news is that the majority of online business offers are mostly business opportunities, where the best benefits are for those who sign up individuals to participate.

Promising financial riches if you join whatever "XYZ" online business hot new opportunity.

Not to mention the enormous amount of coaching and courses available. It's really hard to figure out which online business is the most profitable and least inexpensive to get started that will allow you to make upwards of 20k every single month.

This is what the "Mind-blowing Ways How To Make Money Online In 2020 : Bootstrap Your Way To 20K Monthly" will help you solve and quickly figure out!

No worries this information is timeless and evergreen to take you even beyond 2020 and into the future to discover the best ways for how to make money online and profit 20k or more every month.

How-to Create Information Products That Sell

If I had to think of the easiest and fastest way to make money online, creating an information product would be my first option to make it happen from scratch!

Plus you can get started on a bootstrap budget without spending lots of money.

When it comes to marketing, typically a product is considered to be an object or system made available for consumer use; it is anything that can be offered to a market to satisfy the desire or need of a customer.

How well you market your information product and the demand or need for it in the marketplace will determine the overall value and success to generate revenue from it.

For example in retail, generally products are often referred to as merchandise, and in manufacturing, products are bought as raw materials and then sold as finished goods.

Service is also regarded as a type of product. In general, a product can be classified as tangible or intangible.

Traditional tangible products are physical objects that can be perceived by touches such as a building, vehicle, gadget, or clothing. Where intangible products are products that can only be perceived indirectly such as an car insurance policy.

Services can be broadly classified under intangible products which can be durable or non-durable.

The best information products often fulfill a service need for "how-to" achieve something.

People are deeply willing to pay an insane amount of money for information products that fulfill a service need.

Like instructions, forms preparation, online courses, ebooks that explain how to do something, etc.

Information products are also

membership sites, webinars, live events, special reports, review analysis, exclusive recording, video streams, templates, and more.

And not all information products are always digital and can be offered in non-digital format. Take for example coaching and mentor programs can also be information products, they don't have to always be delivered in digital format.

A growing sector for information products is live in-person events. Taking place as conferences.

This all transformed from the popularity of video courses sold on DVDs and network marketing conferences.

It just takes understanding the methodology of finding the right "information product idea" to develop and launch for major profit.

Step 1 Start Brainstorming

The most significant step to create information products that sell is this first step.

Brainstorm a list of information product ideas.

Search Google, Udemy, ClickBank and other sources to explore topics for information products.

You brainstorming ideas for an information product and seeing what's available in the marketplace similar or like what you want to sell.

Step 2 Choose A Niche

Choosing the niche your information product idea will focus on is the best way to evaluate profitability.

Select a niche concept that is widely searched. This requires doing keyword research to evaluate what people are searching for and the CPC when it comes to cost per click that advertisers are willing to pay towards targeting consumers to search for certain keywords.

Ideally, any niche that is searched over 100,000 times per month or more, will be a good information product idea to launch to make a good profit from.

Identifying the best keywords and SEO options for information product ideas is the secret to success.

Step 3 Decide On Delivery Format

Determine what type of information product you will offer in the marketplace.

Will it be an ebook, course, download, live mentor program, etc?

Source keywords to determine what people are searching for to create an information product idea.

Best of luck!

Dot Crypto Blockchain Domain Investing and Flipping

Have you been hoping to get in early on the next big thing for the Internet??

Well, Dot Crypto Blockchain Domains on Ethereum most highly sought after ecosystem is it.

Buying up early like right now is the best time to get started!

Billions of dollars are being spent on Blockchain technology.

This is an example of a marketplace dedicated to Dot Crypto Blockchain Domains on Ethereum, for sale, rent, flipping, or licensing.

https://uplymedia.com/product-category/crypto-ethereum-blockchain-domain-marketplace/

Businesses are looking for ways to accept bitcoin and consumers want ways to send cryptocurrency!

With Dot Crypto Blockchain Domains on Ethereum, all you need to know is the recipient's blockchain domain to make this happen.

Easily send bitcoin, ethereum, and any other cryptocurrency with just one domain!

That's the next big thing happening on

the Internet, which wasn't made for payment processing went it was first built.

"This is a complete game changer!"

Buying and selling Dot Crypto Blockchain Domains on Ethereum is going to be one of the best ways for how-to make money online in 2020 and sustainable forever.

The problem Dot Crypto Blockchain Domains on Ethereum solve is, no more worrying about sending to the wrong addresses for super long crypto addresses.

Not only that businesses need brand awareness and to build credibility.

The focus of top keywords as Dot Crypto

Blockchain Domains on Ethereum will be the hot ticket towards profitability.

Another great benefit of Dot Crypto Blockchain Domains on Ethereum is these are invaluable crypto assets as domains that are stored in your wallet, just like a cryptocurrency. They can't be moved without the private key owner's permission authority.

Making it possible to point these magical crypto assets domains to content on a decentralized storage network!

Welcome to the "DWeb" decentralized website!

It's Blockchain Domains meet decentralized storage, which will equal big profits for early investors and those smart enough to focus on keywords for Dot Crypto Blockchain Domains on Ethereum.

A key element is that unlike other blockchains, Ethereum can do much more. For example, Ethereum is programmable, which means that developers can use it to build new kinds of applications. Making these decentralized applications (or "dapps") and Dweb gain the benefits of cryptocurrency and blockchain technology.

The marketplace is prime for buying, selling, flipping, and renting just like real estate buy only it's virtual real estate!

Master Affiliate Marketing For Six-Figure Income

Okay, this is a fast and serious way to make money online.

It's almost too simple to understand that most people don't get into it. That's just how simple it is!

"What exactly is affiliate marketing and how can you make money from it?"

Affiliate marketing is the process by which an affiliate (which is you) will earn a commission for marketing another person's or company's products.

Typically, most affiliate programs require no upfront fees or costs. You don't have to pay for inventory, stock products, ship anything, or even handle customer service.

"It's a complete referral situation."

You as the affiliate simply search for a product and then promote the product and earn a piece of the profit from each sale you make!

"Affiliate Marketing" is the Holy Grail of marketing as the granddaddy solution for making money online with a bootstrap budget. You can get started for little next to nothing. You just need your own self-hosted website. Don't try to do affiliate marketing without having your own self-hosted platform.

Follow a few simple steps and you almost guaranteed to make six figures with this practice. You can make money starting out fast and grow more and more over time.

First Step, Is There A Market?

To be successful at affiliate marketing, you need to first identify if there is a market for what you would like to offer.

"Told you it was super simple."

Your purpose is to identify affiliate marketing best opportunities.

Determine if there are actually enough

customers out there who want to buy your affiliate product offer. Identify buying patterns and create affiliate product offers on your self-hosted site that will fill existing demand needs in the marketplace.

Why affiliate marketing is so popular is that you don't have to create your own products!

Second Step, Promote Products You Know About

The most authentic way to promote affiliate marketing products is by having a connection. If you use the products already, then you know about the best features and benefits.

The recommendation will also come across as authentic and trustworthy because you know about the products.

Allowing you to be confident and honest that for reasons to support your recommendations.

Third Step, Create Content That Will Attract and Engage Consumers To Buy

The best way to get people to buy affiliate marketing products online is through content promotions.

Which is why you must have your own self-hosted website to make this happen.

Make your content informative and explain why the affiliate product is your choice and recommendation.

Offering product reviews is the secret weapon to making six-figure income with affiliate marketing products!

The more product reviews and content you have on your self-hosted website, you will be able to create income.

One important factor to remember for affiliate marketing is to follow FTC guidelines for compliance.

As part of its role in protecting consumers from deceptive marketing practices, the FTC requires that affiliate marketers and others using the internet and social media to promote products and brands disclose their financial relationship to the products and brands they mention.

Write a disclosure that says:

"This site participates in affiliate marketing network partnerships. A commission is received when purchases are made from recommendations at no extra charge."

Always place the disclosure before any affiliate links on your page. It's only necessary to place the disclosure only once at the top of your article no matter how many affiliate links are used in the same article.

Secret Sauce To Selling Massive Kindle Downloads

"How I Made 2k, 10k, 100k Self Publishing On Kindle!"

You have seen it, all the people talking about how they made a certain amount of money self-publishing books on Amazon's Kindle, right??

Now, let's break down how you can develop your own online strategy to make whatever you need self-publishing books on Kindle.

While some people will incorrectly tell you that you need to write a book week.

"Who really has time to write one book a week?"

However, if you want to diversify, grow, and expand your Kindle online money making empire having as many books in the marketplace that you can get is ideal. Yet still, you don't necessarily need to write a book every week.

Some top-selling Kindle ebooks are only 6,000 words on average. So, that's not really a lot of words. You can still make thousands of dollars with the right Kindle ebook with less than writing 6,000 words.

"All is needed is to create content in an ebook that addresses a demand that people need to know more about."

There are some 90 million people who are Amazon Prime subscribers and the reach of Amazon's Kindle is massive. You have a greater opportunity to sell ebooks on this platform than anywhere else online.

Now, this should fit in perfectly with your bootstrap budget because to self-publish on Amazon's Kindle is totally 100% free to sell your ebook on the platform.

To make lots of money self-publishing your Kindle ebook you need mainly to understand three top things:

Pricing

The best way to make money is to

understanding pricing for Kindle ebooks.

Starting at between $2.99 to $9.99 is the best-selling price for most Kindle ebooks. Setting your price at this range will net you a royalty of 70% paid by Amazon to you. If you price above $9.99, Amazon pays out a royalty of 35% which will also have to compete with the majority of Kindle ebooks being priced at between $2.99 and $9.99, so please consider that.

To be super successful there is a Kindle pricing secret sauce sweet spot and that price point is $2.99!

This brings you a royalty of $2.09 per each ebook sale. Many major publishers only payout $1.50 on print books priced at $19.99, so you see why people often turn down publishing deals to self-publish on Amazon Kindle?

At the $2.99 price point, you don't have to write over 6,000 words. You can be successful with 2,000 words. As long as the purpose of what the book is about is covered, then people will buy it to save time on what they want to learn more about.

You selling information as an ebook and saving people time by packaging important details for what they want to do inside.

Book Cover

Your book cover is a selling factor for your Kindle ebook.

The more your book cover represents a visual image to relate what the book is about, the better the chances it will be purchased!

If you don't have design skills, you can easily hire people to design your book cover for you as low as $15 on Fivver or use Amazon's free book cover creator tool.

Opt to go for an engaging book title with a cover that features bright colors in bold text. Adding an image that relates to the book subject-matter will attract buyers looking for information on the topic. Making it much easier to sell your ebook in Amazon's Kindle online.

Promotions

Where you will score the most is with your promotions. How well you promote your Kindle ebook is a key factor in building an audience of loyal customers who trust your content and likely will tell others about your book, helping to promote it.

Get the word out about your book by identifying blogs, websites, social media channels, etc that also focus on what your book content is about. These people can help you spread the word and they already have audiences focused on what your book is about expanding your reach.

Establish a self-hosted website for your ebook and link it to your Amazon Kindle page to complete the purchase. On your own site, tell more about the book. Write weekly different blog posts about your book on your own self-hosted website and link to your Amazon Kindle page to complete the purchase.

Don't forget to write a press release announcing your new ebook available and link it to your Amazon Kindle page.

Focus on building an email list on your self-hosted website about your book. This will allow you to inform audiences of new book releases or updates.

If you select Amazon's KDP Select every 90 days you can promote a free five-day promotion or one seven-day countdown deal for your ebook. "Happy publishing!"

Bonus Secrets!

Here are a few of the best ways to make money online for 2020, focus on these niches to make 20k and up monthly.

Blogging

Fitness / Weight Loss

Health

Blockchain Domains Investing

Dating / Relationships

Make Money Online (Why you purchased this book!)

Pets

Beauty Wellness

Baby Products

Self-Improvement

Personal Finance

Cooking

Learn To Play Instrument or How To Sing

Travel

Alternative Spirituality

Investing

- Weddings
- Learn A New Language
- Wealth Building
- Survival Skills
- Green Energy
- Party Supplies
- Tech Gadgets
- Entertainment Reviews
- Sports
- Financing

Good luck and here is to your success!

www.ingramcontent.com/pod-product-compliance
Lightning Source LLC
Chambersburg PA
CBHW070845220526
45466CB00002B/885